1

The Hard Problem Revisited

Fredrik Andersson

ISBN: 978-91-87713-95-8

The Hard Problem Revisited

Summary

An essay on the hard problem of consciousness is a tough task to perform. In what follows I try to cover, if not all, then at least the main problems. *One* such problem is the notion of phenomenal space which differs tremendously from "ordinary" space; the space we live in. I argue that there are several kinds of reality, both the reality of phenomenal space and another reality that pertains to everyday space, only to mention two. Of course this is also valid in lower level explanations such as quantum physics. I assume that these realities intersect in a phenomenologically significant way. As for the hard problem and the explanatory gap I tend to distance myself from semantics and instead I focus on the metaphors we use in everyday language. I argue that even though these metaphors are useful for survival they also lie or give an erroneous image of reality. I also argue that why so many have failed with these problems is that they have, more often than not, neglected ontology. In fact ontology is a key word in this essay. To some extent I touch upon mental causation as it's relevant and I end the essay with quantum physics and an experimental way of doing mathematics mirroring quantum physics. That way we have perturbations from outside and we also have a living system with internal mechanisms that are in constant change. I have by no means exhausted the territory but I believe that I've made a rather solid stand against both idealism and Physicalism. I also touch upon multidisciplinary approaches, most notably Heylighen's, and the role of Enaction in both mental causation, which pertains to the explanatory gap, and also rationality and its converse. In this context I discuss time-space which from the point of view of categories sorts above the other kinds of spaces discussed. The question is if time-space can be fused into the others which, if that is the case, are there for explanatory convenience. In conclusion I discuss mental disorder in the light of Enaction.

1. Introduction

This essay is, perhaps, an essay on one of the hardest issues of all; hence the name "The Hard Problem". Since we're dealing with the mind and different aspects of the mind it is hard to avoid metaphysics to some extent. What then, is metaphysics? It's possible to say that metaphysics is everything that Physics can't explain. The problem here then is that we have no idea what the next genius or so in Physics will find and if that happens then what happens to metaphysics? What happens to Theoretical Physics which is a part of contemporary Physics that can't be tested and hence may be regarded as beliefs? Does it adapt? Does the boundary between metaphysics and Physics change in the light of new empirical knowledge in Physics? What seems safe to say is that there is no sharp distinction between the two and there probably never will be. Dealing with the mind is metaphysics in so far as we bring in theory into the analysis. Example of theory can be physicalism and materialism. In this essay I will try my best in being as unbiased as possible and also mindful about theoretical entities even though that is virtually impossible

2. Introduction to the Hard Problem

This is not an easy essay. In fact, no essay on the hard problem tends to be and that isn't just a kind of word play. The hard problem of consciousness can be approached from a distinction often made in the theory of mind, or the philosophy of mind[1]; we distinguish between access consciousness and phenomenal consciousness. As for access consciousness it is associated with such

---

[1] To me theory of mind is a better conception than philosophy of mind for the sole reason that the word "theory" is pretty straightforward whereas "philosophy" seems quite blurry to a great deal of people. Nevertheless in any topic if there is a meta level it easily becomes philosophy. Given this it makes us able to talk about such diverse subjects as Biology and Chemistry as the philosophy of Biology or the philosophy of Chemistry. It's a matter of how to address the meta level of each of these subjects.

matters as memory, behavior etc and that is not our field of investigation.[2] We are discussing the hard problem of consciousness and that mainly concerns phenomenal consciousness. Why do we distinguish between access consciousness and phenomenal consciousness? The question is reasonable and the answer seems to be that it's practical. I don't know any other reason for the distinction in question and indeed usefulness is as good as anything, especially since we're dealing with such an airy-fairy thing as consciousness,[3] at least according to mainstream science.

In a famous paper[4], Nagel defines the hard problem of consciousness as what it is to be something. This can easily be expanded to, for example, what is it like to see the colour white, what is it like to see a butterfly, what is it like to see an apple tree, what is it like to hear opera and so on. As opposed to this "the easy problem of consciousness" mainly concerns access consciousness and that can be explained in terms of, for example, behavior. The fact that I'm awake, for example, and the fact that I'm having breakfast and so on; all of that can be explained from a behavioral point of view. This can, as far as consciousness is concerned, be understood by way of computation or neural systems.[5] This view however, does not lend itself, at least not easily in any sense, to the hard problem of consciousness which we attribute to phenomenal consciousness. Even such a thing as unusually high neural activity in parts of the brain concerns the easy

---

[2] What all of these things have in common is that they can be successfully modelled. In the case of memory for example there are a great deal of theoretical work that can be modelled and/or simulated in, for example a computer or a similar machine. As for phenomenal consciousness it pertains to first person subjective experience and that assumes, in the case of modelling, a kind of ontogenesis in, for example artificial neuron networks. An ontogenesis under those circumstances may or may not lend itself to what is called autonoetic consciousness which normally is attributed to human beings. Autonoetic consciousness includes such things as awareness of self and interestingly there are cases where elephants are given a paint brush and paint a picture of themselves. I wouldn't suggest that the elephant in question is narcissist, that would be going too far but there is, pretty obviously, an awareness of self in the case of the elephant.

[3] Scientist tend to think that consciousness isn't a scientific field of enquiry. We will hopefully get to that later on.

[4] Nagel (1974)

[5] Neural systems perform by way of electrochemistry. We can also refer to it as neurochemistry which would be a broader term. Neural activity performs in a kind of chemical "soup" where positively or negatively charged chemical entities in principle are let in or out of the neurons. In that way an individual neuron builds up electricity. An action potential is an upper limit of electric charge. We say that neurons "fire" when the action potential is reached and this is the basic way neurons function and interact.

problem whereas the hard problem is all about subjective experience.

Nagel set a definition but he didn't label it the hard problem; that came later. There are philosophers and people in cognitive science, like the Churchlands, who neglect it altogether, saying that it's a pseudo problem.[6] However weird Nagel's definitions seems to be it's a very handy definition of consciousness[7]. However, even though it's a useful definition that makes sense to many there surely are quite a few who don't see a problem here, so what then is the hard problem of consciousness? Following Nagel then if there is something it's like to be a system, living organism, computational entity etc then changing places with another system is something very different from if the lights go out. Hence the little notion of subjective experience[8]. Subjective experience is something that we take for granted in everyday life and since it's so immediately present we, more often than not, do that and it's not unnatural. So what is the hard problem anyway? Some people will question if there really is a problem; isn't it just common sense?

3. Expanding Further

Returning to Nagel's minimal definition that if there is something it is like to be an entity then even if you could change places with it, it wouldn't be the same as the sun rising. This something, whatever it is, that it's like to be an entity is, following Nagel, consciousness, and we don't even have to understand it. Expanding on Nagel's account and putting this into what we've touched upon above then Nagel's account would concern the so called hard problem as distinguished from the easy problem as outlined here earlier. Hence we have phenomenal

---

[6] In her very ambitious book "Neurophilosophy", Patricia Smith Churchland equates mind and brain. This is almost an extreme way of doing philosophy but it does lend itself pretty easily to reduction which is out of this essay's territory.
[7] Or rather Phenomenal consciousness.
[8] We're still following Nagel here.

consciousness and we leave access consciousness to the easy problem.[9]

Why then is the hard problem a hard problem or why is subjective experience[10] a hard problem? Why is it a problem at all? One thing that is often attributed to consciousness is qualia. The notion of qualia[11] is a tool to depict the qualitative aspects of consciousness. But what is this more than just another word? We will suggest that qualia, even though it's a way of talking about qualitative aspects of consciousness, explains very little.[12] The qualitative aspect is in fact what it is like to be so if we introduce a concept such as qualia then we run the obvious risk of oversimplifying the hard problem.[13] On the other hand qualia seems to be a practical concept to address the key point here. One argument against qualia is that qualia isn't computational. Another argument is that qualia escapes normal scientific procedures. What this means is basically that qualia doesn't seem to fit into a scientific framework. This suggests that if we stick with qualia then we have a concept depicting something very elusive, a qualitative aspect of mind, but we're far outside the scientific field of inquiry. If we take visual perception as an example and we do that mainly because visual perception is the most explored mode of our senses then if we call what we see input then we have a pretty good idea as to what happens from the eyes by way of the LGN and further back in the brain to the visual cortex, the spatial neocortex and this can be empirically tested and proven as far as the word proof applies but then something mysterious happens because somewhere we simply lose track of the percepts after all these computations have been made.

---

[9] It's not that the easy problem is all that easy but it differs tremendously from the hard problem.

[10] First person perspective

[11] Where a quale is a subjective experience.

[12] Similar to the distinction between access consciousness and phenomenal consciousness it's perfectly fine to say that qualia is a useful concept. This may be alright in everyday discourse but it seems to say very little other than depicting subjective experience. The question is if "qualia" has any explanatory value and it would do that in everyday life where the hard problem is so intimate and immediate that it isn't regarded as a problem at all. In this essay we think differently. It's hardly arguable that introducing a word or concept like qualia solve problems, it merely makes it possible to talk about inner experience (which is, essentially, how qualia is used).

[13] Even though qualia may lend itself to reduction it has very limited explanatory value.

So what happens to the percept?[14] We don't know. Is there any way a research can be pursued to clarify this? None that we know of. But despite this mystery there is this thing like a subjective experience of visual perception and it does feel like something, it does look like something that may resonate and mysteriously conjure up the magic we tend to call subjective experience.

There seems to be a great deal of neural connectivity between the upper cortical regions and the frontal lobe so in theory that could explain the mystery but let us assume that the visual percepts reach these upper cortical regions only to get further transported to the frontal lobe then horray!, we have found a location but have we found an explanation? The answer should be no. Finding a location is definitely not an explanation because we still don't know what it's like. This means that any so called essentialist questions are ruled out and what it is like to see the colour *red* for example is a variety of an essentialist question; hence the difficulty. Finding a location in the CNS can be sufficient from a functional point of view but it's structural nonsense, or rather it is not sufficient from a structural point of view.

A similar aspect is epistemological in the sense that conscious experience is very immediate and so we simply don't notice it in reflection. This epistemological statement I would suggest is very important and it is a part of what makes the hard problem a hard problem. It is the *ease* of which it operates and the *immediateness* of this operation that make it hard for people to grasp that it is indeed hard. This notion could serve as an extension of the definition set forth by Nagel. Maybe it's possible to summarize the epistemological statement above in a more concise way by saying that the ease of the problem makes it a hard problem. Does this statement involve a paradox? A distinction can be made here between

[14] What we think we know is that the input tends to be computed a great deal of times and eventually synthesized to form an image. At least that is what science has to say about the matter. Our problem is really about the impact of that image and we can argue without difficulty that it is not a matter of location in the brain or the rest of the CNS. Rather it is something very different.

logical paradox and natural paradox. I will leave it as uncertain that we have a logical paradox[15] but as for natural paradox the answer to the question is a decisive no. That would be a mistake since there's a great deal going on in the sentence "The ease of the problem makes it a hard problem". Nevertheless it appears to be a paradox and this appearance may be one reason for people not recognizing the hard problem. What about the ease? As already stated there is an ease in the immediateness and operation of phenomenal consciousness which makes it self-evident and for that very reason neglected but the point of dealing with it at all is to explain it and that is the hard aspect. Access consciousness is another story because it can be explained in terms already known but this is not the case with subjective experience, hence the problem.

4. The Explanatory gap and some extra stuff

The explanatory gap is a concept introduced to depict the lack of explanation between the physiological and the mind. The hard problem is what is in need of explanation in order to "fill" the explanatory gap. This would be a little rough and minimal but it's quite straightforward and we will leave it there. The explanatory gap is in need of an explanation. The hard problem is what needs to be solved in order to bridge the explanatory gap. In psychology there's been a number of attempts. One of these attempt is called Constructivism. Constructivism tells us that the mind is embodied and it differs a great deal from the Cartesian notion about the ghost in the *machine*.[16]

So far we've mainly dealt with mind and brain in terms of organization. We can distinguish between dealing with these in terms of function or structure or both but that is likely to make matters more complex, but then again complexity may

---

[15] It's perhaps an open ended question.
[16] Descartes (2002)

be what we need. Phenomenology is a kind of structuralism and as such, we believe, rather apt at explaining at least many of the puzzles of the mind. We will turn to Phenomenology later.

In his classic work *Critique of Pure Reason* Immanuel Kant dwells at length on the mind. Among other things space and time are considered internal and as such we can label them sense modalities.[17] Apart from Kant there was Rosseau and the vision of the intellectual savage.[18] This visionary person has been proven as non existing and even not possible in an empirical world. Kant's inquiry in the *Critique of Pure Reason* is to a large extent an inquiry in what is possible to know. Basically, according to Kant, there is a priori knowledge and a posteriori knowledge. The latter is empirical and the former is not. Kant's enterprise at large is a mediate path with regards to what was on the agenda at the time. Therefore Kant's *Critique of Pure Reason* is interesting but we won't expand on it further in this text.[19] What is lacking in Kant is a kind of process think and even though Kant makes distinctions between, for example, a priori and a posteriori knowledge the former relies on something internal[20] even in the instances when it's a matter of something empirical. There is also the thing in itself which is a very mysterious notion alien to what phenomenologists such as Lévinas are dealing with. In short, Kant is doing a thorough job in epistemology but not so much in the field of ontology.

5. Why not?

Why not idealism? The question is perfectly sensible since idealism pretty easily lends itself to subjective experience. The general problem with idealism would be

---

[17] And this is indeed what Kant calls them.
[18] Le sauvage intéllectuel
[19] The bounds of sense have also been treated by Wittgenstein in a famous quote.
[20] Time and Space

the external world. We can expand on this and by doing so we run the risk of encountering the subject and object distinction and that isn't really where we want to go. Why is this? Well, we're dealing with the hard problem of consciousness and moreover conscious subjective experience in general and in doing this we need to take certain things for granted such as the external world however it may seem to us[21] otherwise the various solutions to the easy problem or problems don't apply and of course we're not into destroying explanation but rather embracing it. Hence technical idealism is left out; at least to a large extent. Such versions of idealism as Transcendental Idealism may or may not fit into the picture. Rather what we've just addressed is general Idealism.   Constructivism can be an idealism and again it can't. Some of the main constructivists have made quite a job in making it non idealist and also non materialist, materialism being the opposite to idealism.[22]

Often we say that an entity is conscious in a way that is the same as this entity being awake. This makes perfect sense but being awake and sleeping are parts of the easy problem meaning it can be explained by way of computation or neural behavior[23]. The hard problem seems to be part of phenomenal consciousness in the sense that phenomenal consciousness is a great deal more elusive than access consciousness the latter being computational or similar. For example the question what it islike to wake up differs agreat deal from the act or mechanism of waking up. Also, a problem with phenomenal consciousness is that it seems very difficult to reduce to neural activity and similar. Phenomenal consciousness may or may not have its roots in neural activity but as such it is extremely elusive and it is also where we allocate subjective experience. As for matters such as access and reportability of mental states we need to find the computational mechanisms behind them and that can be achieved.[24] However, we can't find the neural or

---

[21] Or, simply "world".
[22] Cf Varela F. J. (1996)
[23] See above
[24] A further problem is "What is a state?" and that is somewhat of a mind twister. Surely we need a rigid

computational mechanisms behind subjective experience or to return to Nagel again we can't find the neural or computational mechanisms behind what it is like to be a particular entity.

As for the brain there is a great deal of talk about organization. This can be broken down into function and structure like we saw earlier. The functional aspect tends to lend itself well to the parts of consciousness that concern the easy problem whereas structure tends to lend itself more to the overall organization of the brain. In fact both are necessary but none is sufficient.[25] Churchland dwells on something called *eliminativist materialism* in a way to unify the sciences of the brain. This may pertain to the easy problems but I doubt that it's possible to eliminate subjective experience or the quality of being, sensing etc. Phenomenal consciousness as such is so hard and elusive that it seems impossible to reduce to neural or computational activity and this being the case phenomenal consciousness can't ultimately be reduced to physics.[26] Arguing against eliminativist materialism wouldn't be the same as arguing against a unified science of the mind or brain, it is rather to argue for the embodiment of mind without eliminativist restrictions. In the light of this we have both noted and abandonded emergence theories of mind, and, I believe, also argued that these theories can or cannot lend themselves to a holistic view of the mind. Returning to functionalist explanations it doesn't really matter how much we can explain in terms of function; the problem persists[27]. For example it is possible to argue that subjective experience is located in a certain part of the brain. This particular part of the brain then, it is supposed, has the function of "housing" subjective experience.

---

definition otherwise the concept "state" loses its grip, so to speak, on matters.

[25] As for structuralism it isn't all that clear that it's useful since it seems a bit fuzzy. However fuzzy it may seem we will discuss that later on.

[26] Physics has a special kind of advantage in that it is the bottom line in the sciences. This means that if Physics encounters something that it can't explain then it is left to Chemistry etc. Ultimately all impossible questions end up in Philosophy.

[27] Chalmers (1995). PP 202-204

What have we done here? We have located subjective experience. Location however, is by no means a matter of explanation since the problem persists. From an epistemological point of view it is rather easy to argue that the hard problem prevails but from an ontological point of view it seems like a mission impossible. An example of ontology is a functional explanation and that is left out since the problem persists. According to Noam Chomsky ontology is missing the point but I would hesitate to think so. Opposed to Churchland I believe that an ontological point of view is crucial if we are to find a unified science of the mind and brain. Churchland would argue against that but this is mainly due to her position as an eliminitivist materialist.[28] In fact mainstream research on this topic makes it, first of all, a matter for epistemology. This includes Kant, Descartes, Wittgenstein only to mention a few. It is quite clear that there is something it is like to be a living entity and however elusive it may be there is ontology involved here even though it's from a first person point of view. I will emphasize that an ontological approach to phenomenal consciousness is very important. If there is something it is like to be then that is a quality and since there is a qualitative aspect then it is fair to say that it exists. If it exists then there is an ontological aspect to phenomenal consciousness in general and the hard problem in particular. Thus it seems evident to say that the qualitative aspect, i.e. subjective experience, exists but what we are facing then isn't *that* it exists but rather *how* it exists.[29]

6. More to Come

We've argued against Churchland to some extent and we've also touched upon functional explanations. We mentioned above that arguing against functional

---

[28] To put it very simple an idealist thinks that the mind is immaterial whereas the materialist thinks that the mind is material. Churchland's position says, in brief, that mind and brain can be equated and that the mind or consciousness is nothing but apseudo problem.

[29] It's possible to argue that I exist and the only thing I can be 100 percent certain of is that my consciousness exists. Everything else is secondary from a first person point of view. One obvious problem here is the time aspect.

explanations isn't necessary and not productive since if we do that then we also argue against functional explanations of the brain and also the mind. The fact that vast parts of the brain can be explained in functional terms can't be disregarded even though many of these explanations lack appeal for the average person. Functional performance can be explained from a neurophysiological or cognitive perspective by specifying a mechanism that performs the function.[30] David Chalmers argues in that way and there is a problem here and that problem is essentially Cartesian. Chalmers may have many points but he tends to be a part of a tradition that originated with Descartes. Moreover there is a "side effect" to this and that's dualism. We think differently. A benefit of Churchlands is that dualism is out of the question but the cons of Churchland's position are hopefully touched upon here in a sufficient way. Returning to functional explanation it's a matter of phenomena being functionally definable that is the key thing here. A functional definition can be neural or, in more abstract cases, computational. It is important to note that a neural mechanism is a very low-level definition whereas computational explanations tend to be more abstract or on a higher level in terms of explanation.

What we wrote above is the case in so called reductionism. To reduce a science to another at a lower level is something very appealing to many and it mainly works due to these higher level explanations. It is arguable if reduction can be performed in full and we're hesitant to say so concerning phenomenal consciousness which is what we're mainly dealing with here due to the hard problem.[31] The functional dilemma is whether it is a sufficient explanation or not. Since we have made a distinction above from the organization of the brain into a functional aspect and a structural aspect the suggested answer would be

---

[30] Cf Chalmers (1995)

[31] To reduce a science to another science at a lower level is a dream that many scientists have. However it entails the use of so called bridge laws that work in an interim making reduction possible. The grand task then is to construct these bridge laws and as far as I know there's been a great deal of dreaming and talking and not any action in the territory. The question would be which kind of scientist would be apt and appropriate to perform such a task and second is it epistemologically possible at all. In theory it's ontologically conceivable but that doesn't do anything.

negative.[32] There seems to be an explanatory gap between these functional aspects and experience or qualitative first person experience and given this then this gap, of which there is a pretty broad consensus, needs to be bridged. As for first person qualitative experience it can be functional but that seems to be missing the point because whatever function we're dealing with it still ends up with a quality aspect. When it comes to the appealing scientific theory of reduction it requires a kind of theoretical work called bridge laws. These can be formulated in various ways but we won't expand on that since it's outside the territory of this essay.

We've talked at some length about functional explanations.[33] We have also argued that functional explanation isn't a sufficient explanation for consciousness as a totality. This means, in general terms, that the conclusions we may draw differ from, for example, eliminitivist materialism.[34] Since function is ruled out then what we seem to need is a structural explanation. We have also argued that reduction doesn't work with phenomenal consciousness as traditionally conceived and this fact seems to make the hard problem a great deal harder. If we turn this way of reasoning upside down so to speak then what we have is something called *emergence*. Roughly emergence theories of the mind state that consciousness emerges from but is not reducible to Neurophysiology. Of course this calls for an explanation as to how consciousness can emerge from a neural basis. In theory it is possible to go from a neural basis to a computational level and explain it from there but the problem would be how the computational level "mirrors" the neural level. Over the years emergence theories of mind have lost their appeal quite a lot and I'd say it's due to

(a) The phenomenal space necessary

---

[32] This would mean that a functional explanation is not sufficient but in need of a structural explanation or that the functions performed need a structure in which they operate. The latter can be understood as "embodiment".
[33] Following Chalmers (1995).
[34] Churchland's point of view. Churchland (1995).

(b) Problems of mental causation

Only to mention two.

As for Chalmers he talks at some length about something called supervenience[35] and makes a distinction between logical and natural supervenience. This is, of course, an attempt to bridge the explanatory gap. Where this seems to lead us is towards a dualism and also a case against reduction of phenomenal consciousness, but let us expand on supervenience. A rough definition is that supervenience is a relation between two sets of properties; B-properties[36] and A-properties. Roughly we can say that B-properties supervene on A-properties if neither the B-properties nor the A-properties differ. This is a formalization of a relation between, let's say Physics and Chemistry. There may be two distinct concepts of mind where one is a phenomenal concept of mind and the other is a psychological concept of mind. The latter is about what it *does* and the former is about what it *feels*. In this respect a structural explanation of qualia seems plausible. However it shouldn't be neither an easy task nor a brief task.

Let us expand further on supervenience. Roughly it's a concept that is to a large extent about dependence. The dominant position is Physicalism or materialism and not only that but nonreductive Physicalism. In the light of this supervenience promises a lot for the Physicalist since it provides Physicalism with a metaphysics that seems to fit happily into the functional framework. Basically supervenience is a relation between two sets of properties, supervenient properties and base properties. Then there is something more etheral and that's mereological supervenience. That runs as follows. Properties of wholes are fixed by the properties and relations that characterize their parts. Then why supervenience?

---

[35] This concept -supervenience - is very tricky to translate. Even native English speakers tend to have problems with an understanding properly. It has been suggested to me that "super" as a prefix would, in combination so that we can make up the word in total, that it is reserved for God because whatever it is it imposes itself from the skies.

[36] B-properties would be the high level properties and A-properties the low level properties.

The mental supervenes on the physical BECAUSE mental properties are second-order functional properties with physical realizers and no non physical realizers.[37]

6.1. More on Supervenience

We've already touched upon a rough and minimal definition of supervenience but it is far from sufficient. First of all let me assert that supervenience is a fashionable concept and its appeal lies primarily in the physicalist/materialist realm. Basically, supervenience conducted properly promises a nonreductive physicalism and that is essentially where we are these days. Kim set the scene back in 1982 when the concept supervenience was introduced but it didn't, as far as I know, become "fashionable" until the 90:ies. Both Chalmers and Kim have written extensively on the subject and here we have only touched upon it a little. Considering its importance for physicalism especially it's worth a paragraph of its own. An argument for phychophysical supervenience runs as follows:

> The mental supervenes on the physical in any two things (objects, events, organisms, persons etc.) exactly alike in all physical properties cannot differ in respect of mental properties. (Kim, 1996, p 10).

We can rephrase this in a more succinct way. *No mental differences without physical difference.* This is hardly a very orthodox way of language but it does say something about the thesis quoted above. Some may argue that supervenience is rather brute and that it (even) misses the point searching in vain for mental properties and that is a way of reasoning that I myself tend to regard as plausible. Regardless there's a whole lot of principles to take into consideration. So much in fact that supervenience as such becomes over determined and I'm not aware of

---

[37] What I just wrote may seem as qualitative nonsense but just think about it.

anyone other than me who has stressed this point. Over determination is exactly what we don't want and the fact that supervenience is prone to this makes it, if not interesting as a thought experiment, then quite useless in any serious explanation of mind versus body. We can, of course, argue in a formalized way here and end up with a positive result but what we are doing then is pointing to one or two principles and to derive how supervenience in fact works is not possible from such a minimal way of reasoning.[38]

Now, this is not the whole picture, luckily, for supervenience. We can make distinctions (which we often do) between weak, strong and global supervenience, each with their own specifics and flavor.

Weak supervenience: In every possible world w necessarily if every couple of objects x and y in the same domain is indiscernible in respect to P-level, x and y are indiscernible in respect to supervenient M-level.

Each one of these three can be formalized in logic but for convenience we omit that formalization even though it has an aura of "truth" over it.

It's worth noting that each of these three lend themselves easily to Modal logic, a field of inquiry we will address later on. As for supervenience I think I have covered enough and so I end this paragraph here.

6.2. Where We are Now

Phenomenal consciousness poses a problem in that it seems impossible to explain it in terms of neural behavior and this would also be the case with the hard

---

[38] There's a whole lot of things going on in mental life so how many principles do we have? 20,000? How many of these fit the supervenience constraints? Ten?

problem. To put it in the terms of Ned Block it's like what pops up from Alladin's lamp after it's been rubbed,[39] and very mysteriously so. The problem here is how is it possible to find a neural basis for the hard problem? We have argued along Chalmers at length earlier and we've asserted that Chalmers is a dualist with respect to the problem of consciousness and in being that he seems to fit into a Cartesian tradition. Dualism has its pros and cons but there are other ways to look at these matters than from those we've touched upon so far. We mentioned earlier that Phenomenology as a form of structuralism can be well suited for explaining the hard problem and we will get to that. In his 1996 book Chalmers dwells at length on something called philosophical zombies (p-zombies) and inverse zombies (I-zombies). He dwells on it so much that he exhausts the territory and little more needs to be written on it. An interesting distinction, however, is the distinction between epistemic gaps and ontological gaps. The concept of qualia fits happily in here. Due to the nature of phenomenal consciousness, which itself is a quality, then we don't need qualia because what we do when we talk about qualia is a meta discussion about, essentially, metaphysics. If we're not into creating a metaphysics of metaphysics, and that would be funny, then there's no need for a concept such as qualia because it's already inherent both ontologically and epistemologically in phenomenal consciousness.

In a famous research report Lettvin et al demonstrates that there are additions to the frog's visual perception. At the time this was a very prominent finding.[40] What does this have to do with the hard problem? As for visual perception it was quite ground breaking in saying that "We're not only dealing with input here but there is, to some extent, an outward going process." This finding is at odds with the predominant view of perception as input representations and as such it is not in line with the Cartesian model. Since it is not in the Cartesian tradition then

---

[39] Cf Block (2002)
[40] Lettvin et al(1959)

theories of emergence may be ruled out by this model but dualism at large certainly is. There is a famous slogan in constructivism that "Cognition is in the eye of the observer" and Lettvin et al became the starting point of something called biological constructivism. Later biological constructivism changed its name to enaction. Enaction is a very prominent research paradigm[41] that is yet to become fashionable in science. There is a great deal of great work in this territory; especially by Humberto Maturana and Frasisco Varela.

Below is a picture on how Varela et al (1996) considers the territory. It is well noted that there are three paradigms here and the people depicted with dots as to where they belong are all classics in their respective territories. The paradigms are called "enactive", "emergence" and "cognitivism" respectively. We can see obvious problems here with, for example Daniel Dennett, who is too much a philosopher perhaps, to be dealing with these issues at large:

---

[41] Or similar

A Fundamental Circularity

Artificial   Intelligence

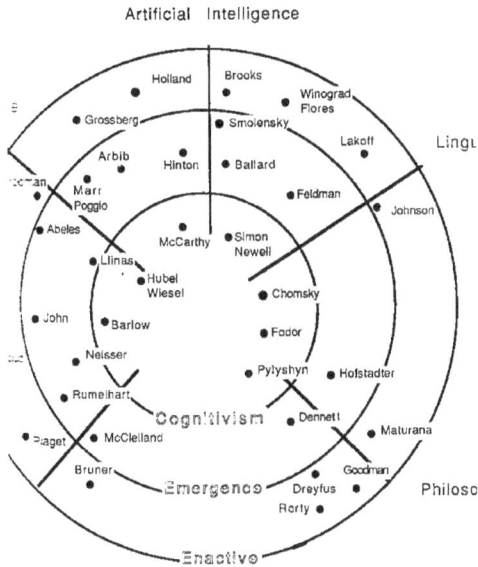

7. A Reply to Michael Tye

In what follows I will address at least some of the problems posed by Tye in his influential book. The upshot will most likely be that Tye misses some serious points and I will argue that he primarily makes two mistakes that don't go together. Tye's book is called *Ten Problems of Consciousness* but I will not address all of them since that will take to much space. It's not written yet but I believe that it's by the end of this section that things start to get real serious.

7.1. The Inverted Spectrum

The argument of the inverted spectrum is strikingly similar to the argument about
the girl living in a black and white box. In the latter case the girl lacks ever having
seen the colour red and hence there is a qualitative aspect missing which we can
call a lack of phenomenal first person experience.[42] The inverted spectrum
argument runs roughly as follows. Assume that Tom has a peculiar visual system.
When he looks at red objects then what is like for him is the same as what it is like
for other people to see green objects and vice versa. Nobody is aware of this
peculiarity. Also Tom has the same linguistic behavior as anybody else. So! When
Tom sees a ripe tomato then his experience is phenomenally and subjectively
different but *functionally* the same as when you and I undergo the same
experience.[43] Why is this? To put it simple the "redness" of the tomato triggers
the same functions in Tom's brain as it would for others to see "greeness". From
the point of view of *function* they are the same although from a phenomenal
point of view there's a big difference.

We've been stressing function here because it's very important in this argument
to see that function does not apply to phenomenal consciousness. We have
already argued that explanations in terms of function can be applied successfully
to large parts of the CNS and more[44] but not to phenomenal consciousness. The
intrasubjective inverted spectrum argument lies behind the latter Wittgenstein in
his *Philosophical Investigations* and the famous private language argument. The
rationale would be that if we can find an intrasubjective inverted spectrum
argument then we can expand that to an intersubjective inverted spectrum
argument and this is beautifully put by Wittgenstein.[45]

---

[42] Meaning there is something it is like to see the colour red. Cf Churchland(1995).
[43] Tye (1995)
[44] Including access consciousness.
[45] Shoemaker (1982).

Now! What's wrong with this picture or rather what's wrong with the inverted spectrum argument as stated above? We've already claimed that there's little wrong with it. What we need, however, is to get serious with materialism.[46] Tye has developed a kind of physicalism concerning colour and colour perception which may attract some and repel others. Talking colour sensations or observations in physicalist terms leads us back to the explanatory gap and I would rather say that if it was ever a problem then Tye seems to me to make it worse and for the very same reasons as stated above. If phenomenal content is to be understood in physicalist terms then what's wrong with *that* picture? Tye seems to go the same pathway as Daniel Dennett in a way although Dennett's account is an extreme one. Dennett's account is similar to Churchland's in the sense that all of them tend to "run over" the problem of phenomenal consciousness. This is essentially what Dennett is doing in *Quining Qualia.* What we're left with is a physicalist approach and we leave that to those who are content with it.[47] What Tye is doing in order to solve the problem of the explanatory gap is basically:

(a) Construing phenomenal concepts as special, relative to physical concepts.

(b) He fails to show that any such concepts can be the case.

In general terms, this is only a way to shift from a problem to another and it doesn't seem reasonable to adapt it since we need less problems to deal with and not more. (a) above is quite mysterious and belongs in a realm of its own having metaphysical implications that we needn't address. (b) however is likely to be a huge problem for Tye and I don't know if he ever addresses that issue.

As for poor Tom in the inverted spectrum argument there is a case for intentionality here that we need to address. It's neurophysiologically clear that correlation is improbable between internal brain activity and external factors.[48]

---

[46] We have already argued quite extensively against idealism but we haven't yet touched upon its opposite, i.e. Materialism even though Churchland et al are firm believers.

[47] We should note that Physicalism is the dominant position these days.

[48] Although this is neurofysiologically clear there's a vast body of research in Neuropsychology simply trying to

For example in Neuroimaging and EEG there's simply nothing in the world external to it that can be safely correlated in the common use of the concept.[49] We can say that the organism is *operationally closed*.[50] There's also something that's more or less intrinsic in any argument and that is *representation*. Returning to the explanatory gap it stems from a certain perspective. A certain way to look on matters and words, where concepts and the like can be delusive. What does it mean for Tom in our argument above to be operationally closed? First of all it is a strong case in favor of being an organism. Second, it somehow doesn't rely on representation but rather perturbation. The entire notion of representation leads us back to a Cartesian view of a mechanical universe.

The inverted spectrum argument relies on there being a linear relationship between first person subjective experience and the external world. The thing is that there doesn't need to be such a relationship and there are numerous examples to show this.[51] We have already mentioned EEG and MRI and indeed it's virtually impossible to do statistical correlations between an event, let's say a sudden spike of electricity in the brain and something that happens in the visual field. MRI is a little more special since what we get is coloured images of the brain where the colour tells us what the amount of blood and thus oxygen a certain part of the brain uses. In these cases it's somewhat easier to do stats and get significant results. From a phenomenological point of view the arguments differ but if we are in the world then what is external to us needn't be external facing the big picture. We are then objects to ourselves and that makes perfect sense. If we are objects to ourselves then we are as much the world as the world is the

make correlations between certain brain activity an cognitive phenomena. This is likely destined to fail; we're not talking physics here.

[49] Let's assume that an external object has a property P. Also let us assume that there's an inner representation that's tracking the property of the external object. If we have a correlation then we can safely say that there's a route for intentionality. This is for the sake of the argument. In clinical experiments there are no such things as perfect correlation between a property P and inner representations. This is very clear in Neuropsychology where significant results are rare (if they exist at all).

[50] Please note that there's a difference here between operational closure and causal closure. The latter especially is relevant to us in this essay.

[51] Biological Constructivism or Enaction is one predominant example where relationships like the ones we're dealing with here are basically nonsense. In Enaction there needn't be any such relationships.

world.[52] If this is the case, and we have argued that it is, then the assumed linear relationship between a subject and an external world disappears and the inverted spectrum argument also disappears or dissolves with it since it's an assumption or even a premise that there is such a thing as an external world. From the point of view of Cognitive Science eye tracking can serve as a good example of this. What eye tracking does is plot the focus of the eyes on an image. This plotting can't be achieved theoretically without an outwards going aspect to visual perception. I would like to stress this as very important and hence the subject as object to itself and the world. Despite this the majority sticks to representation as explanation of visual perception. I would consider this contradictory. It can be intellectual laziness in the sense that easy solutions tend to get more credit than complex solutions and representation is an easy solution, nothing more.

7.1.1. The Inverted Earth Argument; basics

Earlier we discussed the inverted spectrum argument and the upshot was mainly that there is asymmetry between the mental and the physical as we conceive it. In this section we focus on Ned Block's argument about "inverted earth". This argument is a kin but not identical to the inverted spectrum argument and indeed Block uses the Inverted spectrum argument as his starting point. The inverted spectrum argument is, as we see it, not without merit and hence an elaboration is in place. Block (1990) provides us with such an elaboration.

In Block's view the inverted spectrum argument can be stated thus:

"when you and I have experiences that have the intentional content *looking red*, your qualitative content is the same as the qualitative content that I have when

---

[52] Loosely put but to the point.

my experience has the intentional content *looking green* (and we are functionally identical)."[53]

The quote above is all about Block's distinction between intentional and qualitative content, which, in our opinion, seems to be a promising distinction. It also calls for problems for functionalism which is a position that Tye clings firmly to. Block proposes a gap between intentional and qualitative contents of experience and that is the starting point of his own argument.[54] Block states that "if an inverted spectrum is possible, the experiential content that can be expressed in public language are *not* qualitative contents, but rather intentional contents."[55] This is exactly what we are arguing elsewhere in this essay. Block calls all this *the fallacy of intentionalizing qualia.*[56] In short the functionalist approach challenges qualia realism. Block's argument doesn't necessarily rely on if the inverted spectrum obtains or not. His task is something else.

7.1.2. The Inverted Earth Argument; getting Serious

Let us bring to mind the distinction made by Block presented in section 7.1.1. above, i.e. the distinction between intentional and qualitative content. Why do we need to define "red" in terms of a quale that red things are supposed to produce in us when we don't have any good reason to believe that there is something like that? My reply would be that *if* there is something like qualia then it's intrinsic to phenomenal consciousness and hence it can serve in a definition but little else. Block refers to his position on qualia as *quasi-functional.*[57] Personally I would attribute function largely to access consciousness for practical reasons but leave the territory open as for phenomenal consciousness. Practically this means that I

---

[53] Block (1990). P 54.
[54] In practise the distinction suggests that functionalism can only be correct in part.
[55] My italics.
[56] This may be a part of Tye's fallacy as I have chosen to call it. We will get to that later on.
[57] Block (1990). p 58

suggest that qualia can, may or cannot be a shared phenomenon.[58] What I am omitting here are the practical arguments or examples concerning inverted earth and I do that for the sake of brevity. However the fallacy of intentionalizing qualia as Block labels it is based on his distinction. Block argues that talk about, for example, the colour red is a conceptual thing which means that if something that looks red to one but green to another is a strange thing to deal with from the point of view of language since any argument from one is applicable by the other one too. What is reported then is not qualitative content but rather intentional content and this is a big difference. It's actually such a big difference that it almost makes sense. Block states that his own brand of qualia realism is quasi-functional. Why do I write *almost* above? The reader might want to go back a little and review my argument.

7.2. The Problem of the Alien Limb

According to Tye this would be a psychological disorder, actually feeling a limb that the subject don't want to acknowledge as his or her own despite the fact that it is. The limb is alien to the subject in the sense that it doesn't feel real and it doesn't look as being a part of the subject. The questions Tye pose are the following: How can the subject of a feeling himself be involved in this feeling? What is the relationship of the self to the phenomenology of the feeling?

An answer to this would be in what's written above in the questions, i.e. phenomenology. According to Merleau-Ponty[59] the "phantom limb" is a matter of intentionalitity and that is neither physical nor psychological. In this case we can extrapolate Merleau-Ponty and say that intentionality is missing; hence the feeling of the leg being alien. I am aware of cases where people have seemed to

---

[58] Or rather it may, may be or isn't a shared phenomenon.
[59] Merleau-Ponty (1945).

have lost their heads and other things and the same thing applies, namely missing intentionality or intentionality in reverse. This is, of course, a rather simplified version of Merleau-Ponty's work but I think it suffices for dissolving the issue in question.

7.3. Perspectival Subjectivity

According to Tye the problem about perspectival subjectivity is, in the first instance, a problem for the physicalist. It would seem that if you have a pain or an ache for the first time it somehow differs from the second time and so in order to fully grasp what is happening to you you need some prior knowledge; that facilitates. This would mean that if you've gone through it before then you have a kind of perspective or an experiential point of view. Tye goes on to state that phenomenally conscious states are subjective in that fully understanding or grasping them requires adapting experience. This is not the case with physical states. The same thing applies to the famous argument by Jackson about the girl who has never seen a certain colour but knows everything about Optics, Neuroscience etc. Physical states are not perspectival[60] in this sense. According to Tye an Android wouldn't be able to appreciate such a thing as phenomenal consciousness since it is incapable of any feeling or experiencing (in the same sense). With this in mind physicalism, according to Tye, is incomplete.

What Tye is up to above is almost hand in glove with Enaction. However I disagree with Tye on the issue of physical states.[61] There need to be experience involved there too otherwise a person couldn't walk.[62] Plain and simple. This is the case also for the phenomenological body. I'm not sure if we have a mistake in category

---

[60] In philosophy, I doubt that any single philosopher has done more on perspectives and perspectivism than Nietzsche.
[61] If there really are such things as states or if they are processes is beyond the territory here.
[62] Walking is a case of motor control but it is also a case of a learning process. That is to say that the body has a memory of its own and that it can be linked to the CNS is of course obvious.

here since walking can be attributed to access consciousness but the very distinction in theory between access consciousness and phenomenal consciousness is there because of convenience. Nobody knows where one starts and the other takes over and there isn't, as far as I know, any sharp definition on the distinction in question. Perhaps it is safe to say that there is a difference in quality between the two but that doesn't necessarily involve qualia since, as we have argued, qualia is a part of phenomenal consciousness and as such hardly needed as a concept in order to explain anything. Qualia would rather serve, as incorporated, to make up a definition between the two types of consciousness that we are dealing with if there are no qualitative aspects to access consciousness and if and only if so they should differ from the ones in phenomenal consciousness, which in fact more or less intuitively seems to be the case.

7.4. Mechanism

According to Tye neural states as such are not perspecival but phenomenologically conscious states are and this calls for an explanation which, according to Tye, concerns the mechanism that underlies the generation of the latter by the former states. Suppose that a person is connected to a device such that he or she can see his or her own brain. The person can see neurons firing at certain locations when tickled and when not tickled these areas are dormant.

What seems to be a tricky task is in fact, I would argue, the easiest one of them all. Let us distinguish between physical noise and neural noise. In principle neuronal activity, i.e. electro-chemical, works in a way that is much the same as "normal" electric activity which produces a magnetic field. This pertains to electro-chemical electricity too and there is where the neural noise comes in. Rather similar to an old fashioned radio the neural noise produces thought and hence, at least to some extent, cognition. Noise can be regarded as a means or a kind of nourishment for this purpose. This, of course, calls for a sub conscious, which is still in need of explanation even though the term was coined by Freud. If this is a mechanism or not is not for me to say, it seems less mechanic than organic, but I believe that it is a sufficient reply to Tye's problem. Hence it is suggested that the gap between the two states is closed by physical principles valid in an electro-chemical environment.

7.5.   Phenomenal Causation

This is a problem that we have dealt with quite a lot in this essay and are continuing to deal with. The question Tye poses here is if there is a complete physical explanation for why our bodies move the way they do then how come the way seem to us or feel to us can alter this behaviour? The problem is labeled phenomenal causation. Kim has produces a response to this question and so have I in the previous sub section. Nevertheless it is brought up by Tye as yet another of the ten problems.

7.6. Why do the Problems Run Deep?

We have presented five of the ten problems that Tye addresses in his book and two of them seem rather closely related; the problem of perspectival subjectivity and the problem of mechanism. We have also, in addition, presented Ned Block's inverted earth argument since it is related to the inverted spectrum argument.

Tye suggests[63] that we are stuck with a paradox that he calls *the paradox of phenomenal consciousness.* It's not, according to Tye, a formal paradox but it runs

---

[63] Tye (1995) p 37.

like this:

(P1) A man with no hairs is bald

(P2) A single hair never makes the difference between being bald and not being bald (in other words, for any number N, if a man with N hairs is bald then a man with N+1 hairs is bald)

(C) Therefore, a man with a million hairs is bald

According to Tye the premises seem alright but we know that the conclusion is wrong. We think differently. Premise 2 (P2 above) is obviously not true because there must be a line to draw somewhere. Hence N and N+1 doesn't apply. This would mean that only premise 1 holds. Tye thinks that this is a paradox, even though not a formal one but since premise 2 is obviously not true then it isn't even an *informal* paradox but nevertheless it serves as a starting point for Tye.

7.6.1. Expanding on Paradox and Fallacy

Tye asks if the physical world must be objective and the answer to that should be negative, although the problem of perspectival subjectivity that Tye continues to pursue here may not go anywhere. Tye asserts that phenomenally conscious states require a certain experiential point of view. Well, so does the bodily memory. Should we then say that both are perspectival, or better put, should we say that both are intersubjective in nature and not objective in any significant meaning of the word? Tye states that physical items are not perspectival and that is possible in so far as they aren't alive; if alive it would perhaps be a different story. I will now quote Tye as he goes technical:

"In a case of realization, the lower-level type, P, synchronically fixes the higher-level type Q if P realizes Q (in objects of kind K (humans, diamonds etc)), then the tokening of P at any time t by any object O (which is a member of K) necessitates the tokening of Q at t by O but not conversely. What sort of necessity is involved here? It is sometimes supposed that if P realizes Q then, in all *metaphysically possible worlds* every instance of P is an instance of Q. However, I shall adopt a much weaker requirement, namely, if P realizes Q (in objects of kind K), then, in all possible worlds sharing our microphysical laws and microphysical facts, every token of P (in a member of K) is also a token of Q. A second aspect to realization, in my view, is that the determination by the higher-level type by the lower-level type is always mediated by an implementing mechanism."[64]

Let us now rewind for a moment and remember that Tye puts forth a representational theory of consciousness. This is a vast paradigm that is destined to deal with physicalism or the physical. As such it doesn't necessarily deal with the world in any phenomenal sense of the word. This is an important feature to point out regarding representation. It is also a part of *Tye's fallacy* as I may call it. We have touched upon enaction[65] and we will get to that more later on, but enaction would serve as another paradigm within the cognitive disciplines than representation.

Turning to the formalized quote above Tye tries to deal with the physical or physicalism in various ways and the quote may be regarded as an upshot or end result. Consider a whole and its parts. There we have K and O where O can be any object which is a part of the whole, i.e, K. Now the weaker claim that Tye makes in the quote equates any living creature or system with any object. Does that make sense? Maybe. Tye goes on to state something about possible worlds sharing our micro-laws and facts which makes the picture look more appealing but the final remark is that the realization of a higher-level type from a lower-level type requires a mediating mechanism simply does not hold. It may or may not hold for a living organism but I can hardly say that it holds for the rock and ice that

---

[64] Tye (1995) Pp 41-42.
[65] See note 52 above.

constitute a high mountain, just to take a high mountain as example here. The upshot of this would be that P realizes Q in living organisms but not in the world as a totality and this is in line with what I have labeled *Tye's fallacy* above.

8. Conceptual differences, Phenomenal Space and Phenomenal Objects

There are many ways to respond to and handle the explanatory gap and one of them deals with it in terms of conceptual differences. Roughly philosophers can argue that we have a conceptual apparatus that pertains to a physical world and another conceptual framework for dealing with first person subjective experience. The explanatory gap then stems from the difference between these conceptual frameworks. One thing that's worth noticing here is that philosophy is mainly about concepts and little else so in a way the dog bit its own tail here so to speak. We have already touched upon an explanation and that was the subject as object to himself and the world. I'd say that is a strong argument that pertains here too. Just like there needn't be a linear relationship between subject and world as argued in the case of the inverted spectrum argument, need there be a difference in the conceptual frameworks we use for world and self, and this is taken for granted in the argument above. What we have then is a non linear relationship between the subject and the world.

8.1 Phenomenal Space Expanded on

Before we go further here it's important to note that we more or less live by metaphors in everyday life and often without considering them to be - yes - metaphors. One very common metaphor is the container metaphor. If we talk about things that have the container metaphor inherent in speech then what we end up with is a static kind of view of the world.[66] [67]The container metaphor lies behind some things like phenomenal content and phenomenal objects and if we're to examine these in depth then there would be little or nothing left of the contents of mind. We've been treating mind or the mental to some extent but what about mind as a container? Please note that the container in question is basically a metaphor and it's intrinsic in everyday talk. If we are addressing the content of consciousness or mental objects we do it with the container metaphor since it's inherent in everyday talk. I'd say there something fundamentally wrong with this picture. As for Edmund Husserl his method or methodology was all about escaping the *natural attitude* as he called it and reach a sphere of pure intuition[68] and I really don't know what happens with all these metaphors that are inherent in everyday speech but an important thing to realize is that these metaphors are just that - metaphors - and we have to have them otherwise our entire language is at stake. Since we're in the philosophical mood now and since we're talking about a conceptual apparatus then it's relevant to see the metaphors on which these conceptual differences reside. However, this is a grand task that deserves a treatise of its own but let us suppose that one or two metaphors lie beneath all that is said or written about the explanatory gap and also the hard problem then from the point of view of semantics the whole problem seems to vanish.[69] In the real world however, if there is anything that can be regarded as real or *reality* the explanatory gap seems plausible but that would only be the case if we omit any talk about conceptual metaphors and pretend or forget all about these conceptual corner stones which essentially are metaphors to us. The problem seems to be that there cannot be any talk about conceptual differences if there weren't underlying metaphors. To add it up some important features that are inherent in language also makes the explanatory gap

---

[66] The inherent state of the world is opposed to a world in flux and it has its advantages, for example in surviving, but it isn't a true picture, so to speak, of the world as such. The metaphors we live by seems valid for us but basically I'd like to construe them as tools for survival. Please don't mix this up with Darwin or Darwinism which is a different story even though if one merges the two there should be a great deal of significant results.

[68] Anschauung in German.

[69] And please note that we have only made one and only one assumption here concerning language.

and the hard problem to crumble from a philosophical point of view. However, thankfully, there's more to the explanatory gap and the hard problem than just the words. It's easy to accept the correspondence theory of truth and in doing so[70] what we are in fact dealing with is more than "just" semantics.

The argument about conceptual differences seems to be akin to what Tye is doing with colour. Basically Tye is dealing with colour as if phenomenal colours   have properties that are similar to physical colours and this would then, so to speak, close the gap. This is akin to, but not identical with, the argument on conceptual differences.[71] Tye is treating consciousness as a physical phenomenon and a phenomenon that we don't understand. Then he starts to deal with phenomenal experience in general and colour in particular. Colour in materialistic terms however, doesn't seem very appealing because, opposite to Tye's claim, we seem to speed down the lane to Cartesianism and perhaps Platonism real quick.[72] One reason for this is that philosophers pretty often take materialism and physicalism for granted and they do not have or develop a conception of what the material world really is. So far we've been dealing at some length about idealism as a technical improbability but we haven't yet addressed materialism as its technical opposite. One may say that if idealism is not the case then materialism must be the case. If so then what we need is a solid account of materialism but as far as I understand matters there is none. This is not something that we lack; we can rather easily do without both idealism and materialism and this may provide us with a more tangible view on matters.

Returning to Tye and his account of phenomenal objects one thing is clear and

---

[70] There are many different versions of the correspondence theory of truth.
[71] One thing that we haven't touched upon and will not touch upon is treating phenomenal experience as objects. The reason is that it would be a platonic thing to do and few, if any, would take that seriously.
[72] What we are saying here is that dualism persists quite regardless of the nature of phenomenal colour. Treating the mind as material speeds up the process since we haven't, in Tye's description, a clear understanding of materialism. Hence equating the material body with a material mind doesn't appeal because we need a framework for what materialism is all about and as implausible it may seem, we haven't. A spade is a spade and ... So what?

that is that there isn't many people who hold Tye's view. However lack of quantity can hardly be a valid argument.[73] Tye has developed a theory that is mainly physicalist in nature so talking about inner phenomena like colour and talking about the explanatory gap is basically the same thing or we haven't changed the subject. There are quite a few responses to Tye and I'm not going to dwell on that too much but a physicalist view on colour in a physical world can have strange consequences. Earlier we stated that Tye is doing something similar to Dennett but Dennett is an eliminitivist where Tye is a physicalist about consciousness. The explanatory gap as such challenges both versions. We argued above that different kinds of concepts pertain to the everyday world on the one hand and phenomenal consciousness on the other. As for that the upshot would be that the explanatory gap stems from conceptual differences. This suggests that the explanatory gap is about language and nothing else. However appealing I would label it a simplistic explanation. Brie Gerter provides a very elegant argument against Tye. According to Brie, Tye differs from the claim that identities never need explanations because they constitute ultimate explanations; for he allows that identities such as "Water = H_2O" are explainable. Unlike water and H_2O, which are descriptive concepts, phenomenal concepts are "perspectival" and hence irreducible to descriptive concepts according to Tye. In order to make matters harder we have space and time, i.e. "normal" space and time and that is from a third person point of view in an assumingly Newtonian world which would be a special case of Einstein only valid on planet earth and nowhere else. But if we're talking phenomenal space, and this is relevant, it can't be reached from a third person perspective. If we open up a skull then what do we find? Nothing other than what we could expect. The phenomenal space that's assumingly there is nowhere to be found. It seems safe to state that phenomenal space is made up by perception and without perception there can't be such a thing as phenomenal space. What then, are the properties of phenomenal space? Well, it relies on the senses or sense modalities as we prefer to call them. The question is what justifies the concept "phenomenal"

[73] At least not in what we're dealing with in this essay.

in conjunction with "space"? That is a very hard question. Phenomenal space is interior and also experienced whereas what we'd call normal space is exterior but also experienced, only in a different way. There seems to be qualitative differences between phenomenal space and Newtonian space. If phenomenal space is space at all, and we are consistently using the container metaphor here, then what if we didn't think of phenomenal space by way of the container metaphor? This as such makes up a difference in terms of quality because the very word "space" carries with it a container metaphor connotation which is something we live by in everyday life. Let's leave everyday life behind us and focus on phenomenal space as such. What seems intuitively clear is that there is content in phenomenal space but this assumption is challenged from Neuropsychology which is a discipline with quite meager results.[74]

It is clear that phenomenal consciousness is about first person subjective experience but the hazardous thing to notice here can be formulated thus: (a) "What are phenomenal concepts?" and even more strongly (b) "What are phenomenal objects?" Tye would respond to (a) that there is no difference between phenomenal concepts and physical concepts[75] whereas he wouldn't respond to (b) at all since it's outside his territory and the best he can do, I think, is dismiss it.[76] All that Tye really does is shift attention from the explanatory gap to some special nature of phenomenal concepts and then he spends a great deal of effort In trying to explain these concepts despite the fact that the gap is still there. I would call Tye's effort a pseudo problem since if there ever was one then this is it. Tye's view on conceptual differences would render a proposition such as "physical state Q realizes phenomenal state S" impossible to explain for

---

[74] It is refutable that Neuropsychology can find significant results, at least as far as doing statistic correlations between electric spikes in the brain and events in the visual field. And here we go again. I just mentioned "visual field" which is a first person experience whereas the clinical results are there for everyone to see.

[75] Pretty much in line with Dennett, Tye is "Quining Qualia". Tye would say that they have the same properties and that is an idea not worth pursuing since, as we have already stated,there is a huge difference between first person and third person points of view here.

[76] This is what Brouwer does on irrational numbers.According to Brouwer irrational numbers are anomalies and future research will, following Brouwer, come to terms with that subject.

conceptual reasons.[77] This is, of course, obvious nonsense.

A way to formulate the hard problem that differs to some extent from what we've been dealing with so far goes as follows:

How does the brain make up such a thing like consciousness?

Tand (b) need there be phenomenal space?[78] We have already argued that the mind, or consciousness, doesn't necessarily depend on the brain in any other way than through the senses or sense modalities as we choose to call them. This way of reasoning depends on the fact that an organism has senses or sense modalities and that it is through these that consciousness works. This is a very strong claim because it implies that our senses is where the action really is and the senses, or sense modalities, apart from merely playing an important part in experiencing the world and as such are mediators between a first person perspective and a third person perspective also are where consciousness has its roots. To be brief it is the sense modalities that "hang up" the mind and consciousness so there needn't be any roots in the brain for consciousness to rise from. I suppose that this is a clear argument against emergence, however appealing emergence may seem. In the view presented above there is no need to ask "How does the brain produce consciousness" because it doesn't. Our senses, or sense modalities, play an important role as far as conscious experience is concerned, and it's only through these that the mind is realized.he phrasing above is very different from where we started.[79] The hard problem is only hard if the brain somehow makes up consciousness but if it doesn't then there needn't be any hard problem at all. At first sight this seems attractive but even if the brain doesn't make up

---

[77] Or rather for differences in properties and reducibility of concepts. We have only touched upon this argument briefly having given Tye a great deal of attention already.

[78] We haven't dealt with phenomenal space at all and I'm hesitant to start expanding on it since it's a very vast territory that would require or deserve a treatise of its own. Also there virtually nothing written on that topic and that makes it rather tough. It's easier to have texts to relate to but when it comes to phenomenal space it seems like an unmapped territory.

[79] We started with Nagel's definition.

consciousness there must be something like consciousness and regardless of its roots in the brain, as long as there is nothing explanatory the explanatory gap persists. One may then ask oneself if consciousness doesn't have its roots in brain activity then what is the relation between brain and consciousness?

Concerning emergence it has great appeal in that the mind stems from the brain and stretches outwards although it isn't reducible, i.e. It's a one way process and it can't be reversed.[80] Hence it is, in theory, possible to do research on the brain and the emergent mind in emergence is one of those things that aren't reducible back to where it came from in the brain. If what is written above is correct then it follows that we can never find any specific part of the human brain that makes up the mind or conscious experience.

As for what is written above about senses it is important to note that there is a shift in perspective here. This can be demonstrated in that the senses, of course, are themselves rooted in the brain. With this in mind it would seem plausible that the view presented above intersects with emergence to some extent.

Considering the ease of operation and the immediateness of this operation, two factors we mentioned earlier, then as far as phenomenal consciousness is concerned either way it is dependent not on the physical but on perception.[81] What this means is that there are no neural roots to phenomenal consciousness but rather phenomenal consciousness lives a life of its own with regards to Neurophysiology but it is highly connected or linked to cognition in order to exist at all. As far as the container metaphor is concerned its main achievement is that it's useful but since it's wrong[82] and since usefulness is hardly an argument that will last very long then the upshot here is that the mind and thus consciousness or

---

[80] There is a kind of romantic shimmering about emergence that seems to stem from Emerson and Thoreau, both transcendentalists but in a very different way than we would normally call transcendence.
[81] We use   the word "perception" as an umbrella concept in so far as both visual perception, hearing etc are all parts of it.
[82] I think that we've demonstrated to quite some length that it is wrong.

more precisely phenomenal consciousness has a property or two. Being "hung up" on the sense modalities is one property of phenomenal consciousness and this property is hard to underestimate. We have argued that phenomenal content relies on a metaphor but an argument like that seems to treat phenomenal space etc as mere words and as such it is questionable if arguments like that are valid. Let us then assume that there are such things as phenomenal content. If this is the case then just like phenomenal space it is lived experience from a first person point of view. It's very easy to see that phenomenal content is impossible to reach from a third person perspective. Phenomenal objects are even more intricate in that they are metaphysical[83] in nature which may or may not be the case with phenomenal content at large. Phenomenal objects seem to have a metaphysical appeal and similar to Plato's cave where the shadows are mistaken for real, phenomenal object are also mistaken for real. I believe that I've demonstrated the unlikeliness of phenomenal objects as good as it gets. Phenomenal content is another issue because it needn't be objects but it carries with it ontological issues of some importance. Opposed to some in the territory I tend to think of ontology as important in the theory of mind and phenomenal content at large has ontological aspects that we really need to address but that would be a very long treatise that leads us far outside the territory we're examining here. What is clear about phenomenal space is that there's nothing spatial about it and the same thing concerns phenomenal content. It is very hard to talk about phenomenal content without a spatial feature. What we're really saying here is that phenomenal "space" and phenomenal "content" don't obey rules of space and time and since phenomenal "content" doesn't use any space then it's basically the language that is misleading. As for time there seems to be a time as experienced and time from a third person point of view.[84] The upshot of this discussion is that the question how the brain makes up consciousness is also wrong or badly stated since it is clear, from our inquiry so far that the brain

---

[83] If I'm not mistaken the correct concept is over determination.
[84] Immanuel Lévinas explores this beautifully in *Temps et l'Autre*.

needn't make up or conjure up consciousness but from a phenomenological point of view the self is an upshot of its own in that it's basically reiterative and dependent on the senses or sense modalities as we prefer to call them.

When it comes to phenomenal space we've mainly been running in circles here because phenomenal space is primary to the rest, i.e. phenomenal content and phenomenal objects. There is this thing called reality here and it's highly questionable if reality regarding phenomenal space is plausible. If so we need to distinguish between different aspects of reality and these aspects must have a means to intersect. We've seen that we live by metaphors ans when it comes to phenomenal space what we don't need is a metaphor like the one we use in everyday language. This leads us back to phenomenology. What we have is different kinds of reality, some of which are conceivable by human beings. If we take vision as an example it is only possible for us to see in a specific spectrum and that is a constraint on "reality" that we need to have. It is quite clear from what we've written above that phenomenal space isn't space the way we conceive it in our everyday "reality". But suppose that there are other realities. In Modal logic there is something called realism about parallel worlds and maybe that is to take things to far but maybe there is a reality in elementary particle physics that intersects with mundane reality. If this is the case then it's rather easy to argue that the reality of phenomenal space is different from but intersects with mundane reality. One important function of the brain is that it builds up anticipations about the "real" world as we see it. In that way the brain plays tricks with us because we only see fragments of the world but they are very good in the way that they help us survive in the ordinary world.

Chalmers' reply to emergence is that of supervenience.[85] One thing that appears

---

[85] It's worth   emergence theory of consciousness is in line with science at large. Hence it's difficult to argue against emergence in these terms but the mere fact that emergent consciousness has certain propeties, i.e.they stem from the brain and reach outwards has its appeal although it's impossible to find the location in the brain responsible for this magic; and magic it is. By intuition we can disregard emergence since it's in need of a location

as evident is that if consciousness is to be equal to emergence with or without any low level explanation then whatever emergence we are talking about lends itself to a kind of dualism and that is hardly the right way to go ahead. Moreover regardless of the origin of emergent consciousness it is physical in nature and that gives the problem of emergence some extra flavor to it. We are simply not interested in neither (a) dualism nor (b) an emergent property that isn't reducible or in other words that doesn't behave in a two way fashion. In these cases supervenience is by far a more promising route.

8.2. What about Phenomenal Content and Phenomenal Concepts?

We've been dealing at length with phenomenal space so what is left then on the topic of phenomenal concepts? Let us start with phenomenal concepts. One way of looking at phenomenal concepts that seems appealing is to treat them as objects. Frege is probably the only philosopher in his territory that deals with numbers, 1,2,3 etc as if they are objects. This has the disadvantage in the present case of being metaphysics and so we have metaphysics within metaphysics and this would then be a case of over determination. What else can be said about phenomenal concepts? Our discussion above on phenomenal space had the advantage of setting the bounds for what can be regarded as internal and moreover phenomenal space doesn't have any or maybe just a few principles in common with the everyday three dimensional space that we live in and share. One important principle is that phenomenal space isn't stretched out in space and that characteristic alone makes the content of phenomenal space quite different from an everyday 3D computer simulation for example. Sure, in logic it's possible to calculate a3D space but is it real? Whatever phenomenal space is it doesn't

---

and finding a mere location doesn't solve anything. In our field of inquiry location doesn't solve anything. We are asking what it is like and merely finding a location in the CNS doesn't solve anything since the feeling of being would still be there. It may be neurophysiologically relevant but in the big picture of man the location solves nothing. We're looking for a quality and little else.

need much space and this is an important constraint for phenomenal content. I believe that we've already dismissed phenomenal objects, much because of the metaphysics implied but what about phenomenal content at large? We've dismissed phenomenal objects much due to the metaphysics it implies and we have argued that metaphysical over determination is out of the question. Phenomenal content, on the other hand, seems to encompass basically anything that's going on internally and as such it sorts under phenomenal space, the former setting the bounds for the latter. Phenomenal content, however immediate it may seem, is also on a very high level of abstraction. Since it isn't stretched out in space[86] but experienced as such it serves as an enigma for the scientist. In vision, somewhere where the spatial neocortex is located we get this apparent but erroneous image of a three dimensional inner space, and this is stuff that we live by. I'm not saying it's true, I'm only saying that it is something we live by, not dissimilar to the metaphors that build up our ways of communicating.

Phenomenal content can be qualitative by definition since we're discussing something in the realm of phenomenal consciousness which, per definition, is qualitative.[87] Solving this riddle takes us to the next section of this essay.

9. Phenomenology and the Hard Problem

We haven't discussed phenomenology much but rather saved it for later but now it's time to look at phenomenology more closely. Phenomenology is a kind of structuralism and as such, we've claimed earlier, more apt at solving the problem at hand than other methods. In Merleau-Ponty there is constant oscillation both between the individual and the external world but also in the operationally closed individual. This suggests that mind is a product of a recursive function or

---

[86] Or rather time-space.
[87] I've argued successfully against qualia as a concept that we don't need. One argument was that of over determination.

property.[88][89] Lévinas is a similar example. In *time and the other* Lévinas deals with the ego as a hypostasis and that is rather close to Merleau-Ponty's account. In Merleau-Ponty's account intentionality plays an important part in an organism since it works in between the psychological and the physiological.[90] If we have intentionality as described earlier then what about the explanatory gap and the hard problem? The explanatory gap would only be valid if there was a difference between the senses or sense modalities and the brain and this is strange since the sense modalities are part of the brain, at least from a functional point of view. What is the upshot of this? It's too early to tell; we need to cover some more ground in order to make things more clear. As such phenomenology can be regarded as a kind of monism and that is very convenient, in part because monism doesn't lend itself to dualism and dualism is often very hard and also unnecessary to deal with.[91]

We have already touched upon emergence as improbable but appealing. The very thought of local properties nourishing global properties on another level has its pros and cons. Fransisco Varela suggests a rigorous methodology to solve the hard problem. In that respect he is in the phenomenological territory. Rather than talk about function Varela talks about[92] the structure of human experience and this is indeed relevant for phenomenology. As for Chalmers he insist that whatever explanation we have about human experience there is always an extra "ingredience" that is escaping explanation. This is further fueled by others. Penrose argues, using Gödel's theorem, that human thought is uncomputable. Thus human thought belongs in phenomenal consciousness as we have chosen to

---

[88] The concept "recursive" is borrowed from Mathematics and it's the best concept I can think of for describing what is going on. The concept "Operational closure" is often used in Cybernetics and also in Biological Constructivism or Enaction as it's also called. It's plausible that Enaction is close to phenomenology at large and perhaps Merleau-Ponty in particular.

[89] This recursive function works by way of the senses or sense modalities and thus mind needn't be dependent on the brain construing it. It's enough for there being senses or sense modalities and the intentionality will basically do the thing.

[90] Merleau-Ponty refutes both.

[91] There are many different kinds of dualism and monism escapes them all.

[92] Varela (1996)

call it. Human thought is part of phenomenal consciousness. Thus human thought is a case of phenomenal content. An important feature of views like emergence and dualism at large is that the mind or consciousness doesn't have any causal role to play and this is simply very non-empirical. It is tempting to think that if I grasp an object and lift it then there should be some kind of causality between my consciously lifting the object and my thoughts about doing it. Emergence doesn't seem to be able to answer that problem but phenomenology considered as a monism may be able to explain it. It's a very elaborate problem.

One reason why emergence is unlikely is that it lends itself to property dualism as opposed to monism which is what we prefer and phenomenology seems well suited to carry out the task since it emphasizes structure rather than function. Why then is property dualism unlikely? The arguments are already there since it is exactly the kind of dualism that Descartes is into. There are, of course, other kinds of dualism but less successful and more abstract. One example is substance dualism which is hardly around anymore.

Since we've left the Cartesian tradition behind we can safely speak about monism and how does Varela's Neurophenomenology fit that picture? First of all let us suggest that our analysis of phenomenal space makes it possible to equate qualia with phenomenal consciousness and so we have one problem less to handle. It is important to point out that by equating qualia with phenomal consciousness doesn't mean that we are Quining Qualia like Daniel Dennett puts it and a main reason for that is that it is not a physicalist point of view, neither is it a materialist point of view.[93] As for the ontology of the mental it is irreducible and that is a fact that is vastly neglected. The fact that phenomenal consciousness is irreducible seems to lend itself to emergence theories but if so we're in a Cartesian position where phenomenal consciousness is supposed to have local properties that make it up and this is an inquiry that we have already pursued. There isn't any

---

[93] I mention both Physicalism and Materialism here since they are used, more often than not, interchangeably.

distinction between observer and the thing observed and this makes our attempted observation of subjective experience fail. Agreeably, this is a rather special case of epistemological concern or so it seems. Where the ontological issue seems straightforward its epistemological counterpart seems intuitively unclear and why is this? Mainly it's because of the way we organize life and make distinctions even where there are no disctinctions to make without the objection of counter intuitivity.

Varela has a problem though, and that problem is that within Phenomenology there is no unity as to methodology and I doubt that Varela is able to present one. Husserl, often described as the founder of Phenomenology, had a method but this method has obvious flaws in part pointed out by Varela himself. So what about the methodological remedy? Shall we say like Brouwer[94] in the philosophy of Mathematics that it's a problem for future research to solve? By judging from our example above that there isn't any distinction between the observer and the observed but that, as the Enaction slogan states, cognition is in the eye of the observer[95], we have already equated qualia with phenomenal consciousness and as far as a methodology is concerned we are facing a problem with the subject examining itself. What does it find? Nothing at all and this is not surprising since we have a distinction between subject and predicate in language that simply isn't there in the "world." What we must do then is to use process words and if so the "subject" is a part of consciousness if it exists at all as an individual instance. What we are dealing with here as a methodology, I would suggest, is nothing but a reiteration by way of which the human mind operates with the world. We need to address this reiteration in its principally mathematical structure in order to get anywhere. Translating this mathematical term into Phenomenology or Neurophenomenology needn't be a problem since its principal outlay is ready in

[94] Van Atten (2002). It may seem strange that we have a Matematician in Philosophy but if we are to look at the foundations of Mathematics then these foundations tend to become a task primarily for philosophers. Moreover there's more to the picture than Brouwer in this territory. Gödel is another example.
[95] That cognition is in the eye of the observer or broader that cognition is in the sense modalities of the observer clearly suggests that cognition is doing the distinction making.

Mathematics. Hence we needn't prove it in some more metaphysical way; this won't do as Mathematics is mainly an ideal science. As for our discussion above that distinctions essentially are cognitive and that language as is often tricks us mainly due to the distinction between subject and predicate which is a linguistic distinction that needn't correspond to anything.[96] In Neurophenomenology we have the subject examining itself where there needn't be a subject but rather a "stream of consciousness"[97] and if we consider Merleau-Ponty, another prominent phenomenologist, then the subject plays a very limited role if any. Merleau-Ponty is happy off with the reiterative explanation and I would suggest that this is a viable route to first person experience. If we talk about first person examining itself then what we're really into, according to our result so far, is a matter of reiteration as such and this may be the same or it may differ from the overall reiteration of the way the mind operated but since we're trying to consider things as processes rather than static distinctions as language often will have it we have no dilemma in a fusion although at first glance it may be difficult in sorting out what is what.[98] What Varela is saying is that there probably is no solution to the hard problem and the explanatory gap and what we can do is shift the perspective on the issue and this is, assumingly, where his methodology enters the scene. The problem Varela is facing in talking about methodology is that there is no such thing as a consensus or agreement on one methodology and by suggesting Neurophenomenology Varela maybe simply adds yet another one in a field that is rather densely populated and diverges to a great extent.[99][100] As a matter of fact the methodology Varela advocates is Husserl's method and Varela

---

[96] For example a statement such as "The flash flashes"can be reduced to "There is Flash"or "Flashing". The latter would do in some languages; mainly eastern.

[97] Dainton (2000).

[98] This I would suggest is a matter of scientific diligence.

[99] Varela starts his paper with one basic point in Chalmers; first-hand experience is an irreducible field of phenomena. He then states that there is no theoretical "fix" or something "extra to it" that can bridge the gap. This may be true if we do not have a process view on matters but since we've introduced a process view on this very vivid field of inquiry the upshot would perhaps be different.

[100] It may be a paradox as such but in his book "The Embodied Mind" Varela et al are doing a thorough job concluding that Husserl's method is wrong and yet in this recent paper Varela relies on a kind of reduction that is outlined by ... Husserl.

has already refuted Husserl's method in a previous work.[101] However it's quite clear that if the subject examines itself then it finds nothing. Moreover if the subject is examined from a third person point of view then, as we have argued extensively in this essay, there's also nothing to be found. These fact must put constraints on the problem at hand both from the point of view of language and also from the point of view of principles attributed to the human mind. What seems intuitively attractive is that a certain set of principles on the electrochemical level has some impact on a "higher order" set of principles concerning the human mind in general and phenomenal consciousness in particular. I would suggest that being Varela, his enterprise is very bold considering the research field or tradition to which he belongs.[102] It also seems intuitively clear that there are a number of constraints on the principles on the electrochemical level that reflect on constraints on the principles concerning the human mind. Some hard work is needed here in identifying these principles.[103]

Apart from what we've written above it seems obviously clear that we need a new take on subjective experience even though Physics in general seems to have made a clear case in an opposite direction, i.e. that of reduction. Reduction seems attractive to most things with one exception and that is what we have called phenomenal consciousness. What is also clear from our rudimentary explanation of Varela's methodology is that it is perfectly in line with something we wrote earlier, i.e. that the subject is both subject and object to himself. This amounts to experience of self and world even though these linguistic distinctions can be misleading, however it is unlikely so if we are dealing with an operationally closed system which is what a human being really is and that makes perturbations crucial. We have argued above that we have five known senses and these five senses or sense modalities constantly interact and "make up" the mind in general and

---

[101] Varela et al (1993).
[102] The bold thing with Varela, of course, is coming up with yet another methodology even though we can assert that he has some very good points. Also it seems that the methodology that Varela proposes is a Husserlian one and the irony of that is that he has refuted Husserlian methodology elsewhere. Cf Varela et al (1993).
[103] I may sound like a reductionist here but that is beside the point.

phenomenal consciousness in particular. If all of these sense modalities are absent in an operationally closed system, i.e. In this case, a human being, then the human being probably dies because intentionality, a concept that we haven't discussed much but more taken for granted, has no field but internal to the operationally closed system and since it's unable to interact then the subject, if it was ever there, ceases to exist. If there isn't any modality for intentionality to work with then the very basis of intentionality dies. Therefore such a system can't live.[104]

Returning to Varela his promise is rather bold in that he supposedly has a methodological "cure" for the hard problem. His problem though is that it's unlikely to perform a first person kind of investigation. If we expand on that the whole thing relies on reportability of mental states where there are none and the reason for that is simply that there are no mental states, there are mental processes and so reportability fails since it only focuses on one thing in time at a time only to move further. The reality of it all is that phenomenal states aren't constant but rather in flux and we will see why by the end of this essay.

As the reader has probably figured out by now I have a definition of first person subjective experience, or simply put phenomenal consciousness, that is rather different from Nagel's and it goes something like this: *phenomenal consciousness is a part of a living system that has the ability of a reiterate loop interaction.* That's it. There are some rather tricky words in that definition but it isn't as raw or minimal as Nagel's; rather there are two core principles in this definition and they are (a) reiteration and (b) loop. At least one of these two can be understood mathematically and also both indicate processes as opposed to states and the latter especially is very important. It's intuitively clear to ask "where's the subject in this definition?". The answer is that the subject and subjective experience is

---

[104] Somewhere there's a paper by Maturana and Varela describing what is and is not a living system.

either in the reiteration[105] or a product of reiteration.[106] Hence, at face value, we have two different answers here and both of them seem plausible. In order to rule one out and keep the other one we need to combine the recursive aspect with the loop aspect. We can assume that the loop aspect is only valid given one and only one interpretation of reiteration and by examining which one fits or intersects we can single out the interpretation that's valid.

10. Mental Causation and the Hard Problem

As for mental causation it seems very difficult at first glance and yet there are some very promising results in that area.[107] Also there is a great deal of controversies in that particular field of inquiry. Let us first assert that physicalism is different from the physical.[108] A fundamental question here is how is it possible for a conscious human being to think about lifting a glass and the physical fact of his lifting a glass? We can illustrate this with a syllogism which supposedly is a causal argument in favor of Physicalism.

(P1) Conscious mental occurrences have physical effects.

(P2) All physical effects are fully caused by purely *physical* histories.

(C) The physical effects of conscious causes aren't always over determined by distinct causes.

The argument seems very straightforward but the crucial part here is premise 2. Physics as a whole *isn't causally closed* and moreover it doesn't pretend to be.

---

[105] The fact that I don't mention the subject or subjective experience explicitly in the definition is because they are inherent in what is going on. I would agree that it's a third person formulation of a first person dilemma but it's as good as it gets.

[106] Lévinas is treating the subject as a product of something that we can translate into reiteration. Cf Lévinas (1988).

[107] The work of Jaegwon Kim on mental causation is one such promise.

[108] Kim has written an eloquent book titled "Mind in a Physical World" and it's important to note that by judging from the title it is NOT a physicalist account but rather the world is considered as physical and this lends itself a great deal harder to theoretical discourse than the term "physicalist" would.

There's *nothing in Physics that suggests completeness* and to me that is an important feature telling us that Physics is a science and not a world view. Here we have two facts of Physics that are clearly at odds with premise 2 above.[109] Premise 2 *appears* to be true but if we expand on it then we see that there are inherent contradictions involved; (1) completeness or rather the lack of it and (2) a lack of causal closure. It is safe to say that premise 2 is false and it's also safe to say that our reasons for refuting premise 2 intersect or overlap. However, it is possible to argue from our counter proof it's a matter of metaphysics and if so premise 2 applies. However, in that case we have a problem with premise 1 even though we may (or may not) have solved premise 2, and that involves conceiving the mind as metaphysics and also understanding physical effects in the same vein and thus we have further complications if we conceive of the syllogism that way. Also the conclusion must be metaphysical so what happened to the human being in this syllogism?

What we've done above is illustrated the significance of mental causation with or without physicalism.

There is something called supervenient causation. What this means is that a given property can derive a causal role in virtue of its supervenience on a property involved in a causal process. This is tangible in so far as supervenience is a valid way to treat the mind-body problem and so far we have little to argue against that.[110] When it comes to mental causation we have tons of problems and we can only touch upon some in this essay. For the sake of brevity we can mention Gödel's incompleteness theorem in Mathematics. What can that possibly say about mental causation? The answer is simple. No matter how hard we try to solve the matter of mental causation we simply won't succeed in full. This is an

---

[109] Premise 2 involves both causal closure and completeness and however compelling premise 2 may look these two inherent features seem to undermine the premise. What we need to do is remake premise 2 maybe add more premises.
[110] Suffice to say that Supervenience is a very fashionable concept and has been since the 90-ies. Hence it's possible to decline in popularity at some point and be replaces by other concepts.

applied formulation. That we can't solve the issue of mental causation in full may be true, most likely so, but it doesn't mean that we need to give up the entire project.

I am not saying that we should give up mental causation. What I'm saying is that it's extremely complex and also too complex for this essay unless I want to write 200 pages and I don't. Thus mental causation is only touched upon briefly here as a part of the whole problem space. I don't intent to try to find a solution; I think there are others who are more apt at doing that, but it does indeed belong among the problems we are discussing here and however brief this section is I hope the reader has received some of its flavor. There is, of course, this thing called supervenience and we have dwelled on that quite a lot. I assume that the reader has a fair understanding about supervenience. It's not only physicalists who appreciate it, there are many other trends and positions that are equally happy. What does supervenience say on such a complex matter as mental causation. For starters supervenience is a way to formalize matters and it entails a kind of dependence which means that it can only operate one way. In part this is included in non-reduction of phenomenal consciousness and many physicalists seem happy off with a kind of non-reductive physicalism.[111] We've dealt with three different kinds of supervenience here and a forth would be called local supervenience.[112] Basically it goes like this: M properties supervene locally on P-properties if P-properties of an individual determine M properties of that individual. Given what I just wrote it would be interesting to compare weak and strong supervenience; especially in the light of modal logic.

Let us be clear about mental causation. There isn't just one but several different problems of mental causation and I'll list them here below:

[111] At least this is a dominant trend.
[112] It's in this territory we find David Chalmers and others. Cf Chalmers (1996).

(a) The Problem of Anomalous Mental Properties

(b) The Problem of extrinsic Mental Properties.

(c) The Problem of causal Exclusion.[113]

In what follows I'll try to deal with these three in order to lay barren the problem space that I believe is inherent in these problems.

(a) says, roughly, that there are no causal laws connecting the physical, and nor are there any laws that connect the mental with the mental. This is referred to as mental anomalous properties. It was put forth by Davidson and, if I'm not mistaken, it's his most prominent work to date. This position would regard mental phenomena as illusions and as a consequence of that it stands out with respect to the explanatory gap.

(b) is rather different. As a matter of fact we have already dealt with it when we discussed phenomenal space and its sub concepts. Roughly it says that it is syntactic structure that is responsible for causation and not the semantic content. However, one problem with this is that there seems to be an anomalous fragmentation of the mental thus giving us, not only an assymetric view on the matter but rather disintegration at worst.

(c) The problem of causal exclusion is basically a problem concerning causal closure. We have previously stated that operational closure and causal closure are two very different things. A living system must be operationally closed but there is no such constraint as to the mental Gödel's incompleteness theorem works wonders with causal closure and hence there needn't be any causal closure which is something that (c) above is dependent on. Hence it is plausible to refute (c).

[113] Kim (1996)

11. Mapping Physics; The fundamental Level dealt with

A view on the matter that seems to promise a solution stems from physics. A fundamental building brick of virtually everything is elementary particle physics.[114] We can suppose that elementary particles are arranged in certain ways within an operationally closed system, i.e. a living organism. Perturbations tend to change the operationally closed system[115]. What this means is that the particles, systemized or arranged in a certain way changes. The interesting thing here is that the particle structure can be translated into mathematical functions and what we have then is a mathematical pattern within an organism that changes or rearranges due to perturbations. Also, from the point of view of biology there can be patterns that are mathematical in nature and what that can mean is that mathematical patterns basically make up or set the boundaries of a living organism. It's worth noticing that we're not dealing with a functionalist explanation here even though mathematical functions can be a computational basis for mathematical patterns. If we take visual perception as an example, and the reason for this is the vast body of research on visual perception, then if cognition is in the eye of the observer[116] what we get is pattern recognition in those instances that the operationally closed system or organism inhibits a mathematical structure that is capable of dealing with pattern recognition. It is safe to say that the human brain arranges itself in order to deal with the world. This is to say that the human brain arranges a pattern that is equivalent to what the human brain expects the external world to be.

I should stress that I'm not a reductionist but interesting things happen if we

---

[114] This is a very low level aspect of what we're dealing with here
[115] This would be a valid argument from the point of view of cybernetics although we may shift from one level to another but that has to be proven as a valid counter argument.
[116] And that is a slogan in biological constructivism.

leave electrochemistry and step down to a lower level which in fact is as low as it goes and that is Quantum Physics or Quantum Mechanics. From a mathematical point of view there are Quanta that are continuous and there are quanta that are discreet, hence the name. Also, in the weird world of elementary particles we can choose which kind of property we want for motion, or rather it is the kind of motion displayed by the particles that tell us which mathematical formulas are applicable. Heisenberg's version deals with a matrix and that is complex and difficult even though the mathematical aspect is clear. Another way is to have particles move as waves and there's an equation, Schrödinger's equation, that rules the behavior of the waves. So, what we have is quanta of particles that move in certain ways either continuously or discreetly. It should be possible to form a meta equation in order to translate what happens on this physical level into mathematical terms. The upshot of this is a mathematical structure and this structure shapes the neurochemistry of the CNS.[117] An important feature of a mathematical structure or pattern like this is that the pattern is in constant motion since what is mapped is elementary particles and they have a tendency to be in motion. Mapping all of this with mathematics provides us with an internal clue to perturbation and we also have a clue as to how the brain shapes itself in order to fit in with the world external to it. The clue tells us that we only know what we're inclined to knowing, whether we want it or not. Are we talking determinism here? The answer should be negative since we are dealing with an operationally closed system viable of perturbations. Any kind of perturbation should change the mathematical structure and also the structure of the brain.[118] The mathematical pattern, on the other hand, can change due to differences or changes at the quantum level. This change reflects in changes of the mathematical pattern in an outward going way. What we have here is really

---

[117] It's important to note that the mathematical translation is on the same explanatory level as Quantum mechanics. It's just another way of handling the changing Physical nature of elementary particles. We will not dwell on the actual Mathematics here since it's very complex and maybe too complex for an essay like this. Mathematical implications and elaborations are hence left for further essays, dealing with this specific specre.
[118] Basically the entire central nervous system can be perturbed and the changes will reflect on the mathematical structure which should rearrange.

perturbations from two parts; one is inwards going and the other, the latter one, is outwards going. If we take visual perception as an example then cognitive penetration is quite possible. What this means in practical terms is that there is (a) a pattern recognition in vision and (b) that there is a cognitive penetration, i.e. the brain is structured in such a way that it fills in patterns that are outwards going to the "image" that is inwards going. The upshot of this is that such things as expectations and other things govern what we see to some extent.[119]

It should be noted that translating the behavior of Quantum Physics into a mathematical pattern doesn't mean that we shift to another level of explanation. Rather it is a *translation* into mathematics that we undertake and the upshot of this *translation* is a mathematical pattern which, at best, mirrors what is going on in Quantum Physics; the fundamentals of a living system or organism. We aren't dealing with another, higher, level of explanation until we start talking about chemistry, or rather electrochemistry in the case of the human brain.

12. The Matrix Problem

When I started writing the section on Quantum Physics and phenomenal consciousness I thought that I was rather original but there already is some work done in that area and so the best I can do is not to copy what has already been written.

Some people who are suggested to be geniuses of the time have addressed the issue if all we experience isn't just a big simulation and as a consequence of that nothing is really "real". Assumptions of a superior species have been made by some very respectful people who seem to suffer from a temporal illness of some sort. I would propose that we needn't postulate any superior species even though

---

[119] Obvious examples of this is a psychosis.

there's a great deal of fx on the subject. From the point of view of logic[120] there is a conception of realism of parallel worlds. Realism of parallel worlds is a problematic position since there are infinitely many parallel worlds and that is simply not plausible. In Mathematics there is a thing called a matrix which, rudimentary, is a kind of coordinate system. We call an axis a vector and a two dimensional vector space is what we have and here every dot in the two dimensional vector has two values setting the position of the dot. A third vector can be added and that makes up a three dimensional vector space. In theory we can calculate as many dimensions as we like and the number of vectors determine the number of coordinates we need for the positioning of each dot. At present it is possible to visualize four dimensions.[121] A matrix in Mathematics is a simple method of describing a vector space and as such it is open to speculation. It is worth considering that mathematics is an ideal science and as such very abstract on its own.[122] The abstractions of mathematics perhaps make it plausible, in the present case, to talk about representation. We can say that a matrix of a certain kind represents a certain vector space. However, what does this mean? Is the matrix on a meta level relative to the vector space? In a way no because everything given in the matrix corresponds to the vector space in question and thus in this abstract case we needn't talk about representation on different levels. It is then safe to say that there is correlation but not necessarily representation since the matrix can be regarded as a sum the figures of which correspond to certain properties of the vector space.

I've baptized this section "The Matrix Problem" since it stems naturally from a discussion of Quantum Physics. I chose to address Heisenberg and not Schrödinger and I did that for a reason. It seems clear that if phenomenal

---

[120] Or rather the philosophy of logic and thus we're up one meta level.

[121] That sounds odd, right.

[122] It's worth noting that the movie "The Matrix" poses issues that concern realism of parallel worlds although in order to "fit in" to the movie there has to be constraints as to the number of parallel worlds that are possible despite the depth of the rabbit hole. I'm, of course, alluding to the part of the movie where the lead character is to choose between a blue and a red pill. The movie as such lends itself very easily to philosophical arguments blended with mathematical rigor. One odd question that seems appealing is "is the red pill really red?"

consciousness is a fact then Schrödinger's equation would be in trouble, hence my choice of Heisenberg's model.   Essentially Schrödinger's equation provides us with wave patterns that are probabilities and as such it is a very abstract mathematical construct that tells us how likely it is for a particle to be at a certain location etc. Heisenberg's model uses the right kind of Mathematics to deal with a kind of motion "pattern" that elementary particles can display. Basically there isn't any consensus about a single model so hence we have the benefit of choosing which model fits the whole picture the best; hence my choice of Heisenberg.

To quite some extent I've talked around and referred to David Chalmers and one reason for that is not mainly his position but rather his presence on the contemporary philosophical field. A recent talk of his deals with the wave theory of Quantum Physics on the one hand and subjective experience on the other. Some key elements are what is the place and role in a physical world of consciousness and is there a reality at the level of Quantum Physics that somehow links the two together.[123] Listening to David Chalmers is interesting because he starts with Schrödinger's equation which is appropriate in his case and after calculating probabilities due to amplitudes that correspond to locations the scientist decides a certain point as the most probable. After the measurement has been made then the particular amplitude/location becomes central and there's a huge gathering around it. This is what is referred to as a collapse. What we have here basically is Heisenberg's model and Heisenberg's model says that "yes indeed; the observer makes a difference" and so do we have a wave function or not? Well, not necessarily because we're outside the territory of Schrödinger's equation and thus the wave function needn't be applicable. However there are ways of doing calculations with Heisenberg also, it's just that we aren't dealing with wave motion but with continuous or discreet motion of quanta and quanta is exactly the end result of our discussion on collapse. Heisenberg's model roughly

[123] Youtube: David Chalmers Does Consciousness Collapse the Wave Function.

states that a certain particle changes its properties due to the observer and this gives us an indication of how much subjectivity matters even at the micro- and nano levels.

What we have then in terms of motion is a mix between continuous and discreet motion and the calculations in Heisenberg's model are often made by way of matrices.[124] In this kind of mathematical environment we have scalar, vector and tensor operators that together, mathematically, "make up" these particles or rather these are principal properties[125] that we can identify in a micro-environment. However, getting into the details of Quantum Physics is beyond the territory that we're dealing with here; suffice to say that the Mathematics is complex and that the three operators we mentioned above can be regarded as properties and that may be strange considering for example a vector which basically is a kind of indicator of force and direction[126] but nevertheless a very important mathematical component. Heisenberg captures something we can call "The Observer Effect" and that has quite an impact on elementary particle physics to the extent that what we consider real may not be real at all.[127][128] Intersubjective agreement isn't objective truth and an important reason why there is intersubjective agreement is because, as human beings, share something in terms of a perceptual system. This fact is highlighted in Heisenberg's model. From the point of view of elementary particle Physics we can have two

---

[124] In vector algebra it's quite possible to calculate at infinitely many coordinates and a corresponding amount of axises (or vectors). A simple coordinate system is nothing but a two dimensional vector space. We can add as many dimensions as we like even though we can only perceive three; doing the mathematics we can calculate on, in theory, an infinite amount of dimensions.

[125] A Tensor operator is a little special since it's more generalized than the others and thus if the others are principal then the Tensor may or may not be principal; it's mainly a matter of how general and complex we want to describe the entities we're dealing with. Basically a Tensor operator is itself a kind of vector, only more generalized. In her book, Churchland (1995) dwells at some length on something called Tensor networks. I assume that it is a kind of NTN (Neural Tensor Network) that she is expanding on.

[126] As opposed to the wave structure in Schrödinger's equation. As for Schrödinger it's a bit weird because we have input to the equation and that input is chosen by the physician. Moreover the input governs the calculations and so the produced wave structure (of probabilities) is a product of initially set variables. Mathematics is supposedly "clean" but without any doubt there is an element of subjectivity in the choice of how the variables are initially set or fixed.

[127] Reality is probably one of the oldest philosophical problems of all. The concept "atom" is attributed to Democrite if I'm not mistaken and in Greek it means "undividable". History has shown us that a living hell can be the consequence of diverging from that meaning.

[128] Please don't go into a store and ask "are these noodles noodles?". People can get funny ideas.

observers in rotation and both share the same perceptual interface. That way Heisenberg's model can be elaborated in an artificial environment such as a lab and the measurements of choice can be performed with a higher level of probability than would be the case with a single observer.

13. Concluding remarks

Rather little is done on the ontology of perception, at least in a straightforward manner and in the west, despite it being able to, in theory, explain a great deal. Ontology of perception, or rather ontology of visual perception, seems to be exclusive to what is referred to as Biological Constructivism or Enaction as it is also called. It is noticable that Enaction deals with visual perception and there is a known slogan saying that "Cognition is in the eye of the observer". Noam Chomsky once stated that as for the mind questions about ontology is beside the point. We think differently. As a matter of fact ontology may be the key issue if we are to go anywhere in this topic.[129]

What we have here as a starting point is Francis Heylighen's very elegant demonstration of a taxonomy concerning rationality and irrationality in visual perception from the point of view of what we would consider to be more Cybernetics than Psychology, even if the two may intersect. This is done by way of regarding the processes at work here as *distinction conservation* in various degrees. I label it elegant because it is done in these topics but it appears to be as valid a proof or evidence as almost anything in Physics. That's the beauty of Heylighen's work. Although it concerns only visual perception I believe that it can be generalized to concern all the five known senses (or sense modalities as I prefer to technically call them). Also I can hardly think of a better way of

---

[129] Cf Lettvin et al (1959).

approaching causuality than what Heylighen does.[130] As opposed to representation Enaction deals with something that I would prefer to call intersection, i.e. mind and world are not exclusively separated, which is the case in representation. With that in mind it seems safe to say that Enaction is a paradigm that differs from what I would call the representation paradigm.[131]

We have covered a great deal of ground in this essay and I believe that the more important upshots have been Tye's fallacy, our treatise on matrix and also Physics. There is, of course, a great deal more to add and since that is the case I can only hope that others will find this essay worthwhile studying.

---

[130] Heylighen (1989)
[131] The latter has dominated the territory since it first begun.

References

Van Atten, M. (2002). On Brouwer. *Wadsworth Philosophers Series.*

Bayne, T. (2004). Closing the Gap? Some questions for Neurophenomenology. *Phenomenology and the Cognitive Sciences* 3: 349-364 Kluwer Academic Publishers.

Block, N. (1990). Inverted Earth. *Philosophical Perspectives, Vol 4, Issue Action Theory and Philosophy of Mind, 53-79*

Block, N. (1999). The Harder Problem of Consciousness. New York University

Chalmers, D. (1995). Facing Up to the Problem of Consciousness. *Journal of Consciousness Studies, 2:200-219*

Chalmers, D. (1996). The Conscious Mind. *Oxford University Press.*

Chalmers, D. (2006). Phenomenal Concepts and the Explanatory Gap. *Phenomenal Concepts and Phenomenal Knowledge: New Essays on Consciousness and Physicalism.* Oxford University Press.

Churchland, P. S. (1994). Can Neurobiology Teach Us Anything about Consciousness? *Proceedings and Addresses of the American Philosophical Association, Vol. 67, No. 4 (Jan. 1994) pp. 23-40*

Churchland, P.S. (1995). Neurophilosophy. *The MIT Press, Cambridge, Massachussetts.*

Dainton, B. (2000). Stream of Consciousness: Unity and Continuity in Conscious Experience. *London: Routledge*

Dennett, D. (1990). Quining Qualia. *Mind and Cognition: A Reader, MIT Press*

Descartes, R. (2002).

Gertler, Brie. (2001). The Explanatory Gap is Not an Illusion: Reply to Michael Tye. In *Mind, Vol 110, 439. Oxford University Press.*

Heylighen, F. (1989). Causality as Distinction Conservation. A theory of Predictability, Reversibility, and Time order, in *Cybernetics and Systems: An International Journal, 1989.*

Heylighen, F. (1992). Non-rational Cognitive Processes as Changes in Distinctions. *New Perspectives on Cybernetics: Self Organization, Autonony and Connectionism.* G. Van de Vijver (Ed.) Kluwer Academic Publichers. Dortrecht

Kant, I (1993). Critique of Pure Reason. *MacMillan Press Ltd. London*

Kim, J. (2000). Mind in a Physical World. *MIT Press, Cambridge Massachussets*

Lettvin, JY, Maturana, H, McCullogh, (1959). What the Frog's Eye tells the Frog's Brain. *Proceedings of the Aristotelian Society.*

Lévinas, I, (1988). Le Temps et l'Autre. *Éditions Gallimard, France*

Mashour, G. A , LaRock, E. (2008). Inverse Zombies, Anesthesia Awareness and the Hard Problem of Unconsciousness. *Consciousness and Cognition (2008),* doi:10.1016/j.concog.2008.06.004

Merleau-Ponty, M. (1945). Phénoménologie de la Pérception. *Éditions Gallimard, France*

Nagel, T. (1974). What is it Like to Be a Bat? In *The Philosophical Review LXXXIII 4, Pp 435-50*

Shoemaker, S. (1982). The Inverted Spectrum. *The Journal of Philosophy.Vol 79, No 7*

Tye, M. (1995). Ten Problems of Consciousness.

Tye, M. (1998). Consciousness, Color and Content.

Tye, M. (1999). Phenomenal Consciousness: The Explanatory Gap as a Cognitive Illusion. *Mind, 108,* pp. 705-25.

Varela, F, Thompson, E., Rosch, E. (1993). The Embodied Mind - Cognitive Science and Human Experience. *MIT Press, Cambridge Massachusetts.*

Varela, F. J. (1996). Neurophenomenology: A Methodological remedy to the hard problem. *Journal of Consciousness Studies 3 (4): 330-349.*

Zander, F. (2000). Early Vision and Cognitive Penetration - a Dual Experiment Approach. *Department of Neuropsychology, Lund University, Lund, Sweden.*

.

www.ingramcontent.com/pod-product-compliance
Lightning Source LLC
Chambersburg PA
CBHW031814190326
41518CB00006B/338